A STUDY OF NUMBERS

A Guide to the Constant Creation of the Universe

A STUDY OF NUMBERS

A Guide to the Constant Creation of the Universe

R. A. Schwaller de Lubicz

Translated from the French
by Christopher Bamford

Inner Traditions International
Rochester, Vermont

Inner Traditions International
One Park Street
Rochester, Vermont 05767
www.innertraditions.com

First U.S. edition 1986

A Study of Numbers was first published in French under the title *Études sur les Nombres* by Librairie de l'Art Indépendant.

LIBRARY OF CONGRESS CATALOGING-IN-PUBLICATION DATA
Schwaller de Lubicz, R. A.
A study of numbers.

Translation of: Études sur les nombres. 1917.
1. Symbolism of numbers. I. Title
BF1623.P9S37 1986 133.3'35 86-7324
ISBN 978-089281112-0
ISBN 0-89281-112-9

Printed and bound in the United States of America.

20 19 18 17 16

Designed and produced by Studio 31
Typography by Royal Type

Contents

Introduction

"It is interesting to note how the world, especially the scientific world, likes to believe in things stripped of all meaning, while other more meaningful and logical explanations are rejected as fantasies or hallucinations.

"What intellectual occupation, more than any other, provides the best proof of what I have just said? None other than science, which ought to make use of the greatest precision in its definitions.

"Science leads all progress, fecundates every activity, nourishes all humanity; and this same science dilates upon subjects that are absolutely of the first importance. It is the field in which it is the easiest to make mistakes—mistakes that have repercussions in every aspect of life and can retard culture for an entire century. This is a terrible responsibility since the consequences are suffered by numerous generations.

"With what, then, do we reproach science?

"With its conservativism. There is its error!

"The materialist conception of our age impedes all progress. The many discoveries continually being made, especially those made in the last thirty years, are no proof of the value of our age's science but—since generally speaking these discoveries are the result of factors other than those given by science—a proof that it divagates and digresses upon a constantly moving and changing wave.

"Ampère would have discovered the electric motor *had he had the idea* of making the magnetic poles of his equipment turn, but *he did not have the idea*, and therefore years passed before the discovery was made.

"Why didn't he have the idea?

"Because he did not know the force with which he was working: this is the secret that prevents science from being truly a science, i.e., 'knowledge.' Science today is only the embryo of a science: namely, the classification of a mass of observations. For thousands of years human beings knew perfectly well that great science compared to which today's science is in its infancy. Naturally, however, it is more logical to believe in the hypothesis of the ether as a body more elastic than steel and less dense than the lightest gas (hydrogen) than to believe proven and verifiable things revealed to us by our history.

"We limit ourselves to the given facts, which are often not verified. These are taught and, since it is easier to find students who believe what the professor says than students who doubt, and since it is more difficult to find intuitive persons who have the soul of researchers (those who prefer to die of hunger following their path, ostracized from all intellectual classes and scientific associations) than to find young people who seek a position in life by becoming Ph.D's, etc.—that is why science has so often remained stationary.

"When Franklin described his theory of the lightning conductor to the British Association, people split their sides laughing.

"Stranger still, it was precisely in England that the lightning conductor had its first great success.

"Does this prove that science, after verification, accepts all truths?

"No. It only shows that science has always been too sceptical and, in the above example, it is simply *conviction by a positive and tangible fact*, the efficacity of the lightning conductor, that won the opinion.

"Nothing has changed since then, and scientists are still in the same state of mind; only now they have invented a whole new set of hypotheses, a theory of vibrations, an electromagnetic theory— that is all the progress.

"Vibrations: here is another theory that takes in everyone and that no one understands. Someone observed sound vibrations, and a dreamer, tossing pebbles into a bowl of water, found himself fascinated by the circular waves that formed there. This became the foundation of a theory that now amazes the world and serves as the foundation for all present explanations, or rather, could serve as their foundation—since, in fact, it explains nothing at all. Science will say that this is enough for the moment, and that what follows next will demonstrate the truth well enough—but we are perhaps allowed to doubt whether it is enough.

"A masterpiece is judged on the basis of its details.

"Red and green spots, forming heads, bodies and arms, or trees, do not by themselves constitute a painting. To make a masterpiece you need the details of the method used to reproduce the object or idea and *the hand of the artist*, that indefinable thing. So it is too with science.

"To explain the sensation of hearing, it is not enough to say that it is sound waves that strike the tympanum of the ear, since following the influence of the wave upon the auditory mechanism, the nerve transmits the sensation to the brain. *His jacet lepus*! Here science no longer explains anything, for the comparison of the nerve to a telegraph wire explains nothing, especially now that we know of wire-less telegraphy! Likewise with the other senses, and with all theories that follow from the theory of vibrations.

"The classification of observations which (as stated above) has been the task to which science has limited itself until now is certainly useful, but it is by no means sufficient.

"Observations, such as those provided by the Crooks tube, which emits rays called cathodic; radioactivity, which makes the ambiant air conductive; emanations designated by the letters ∂βγ—all this explains nothing to a human being who asks concerning the "Why?" of these phenomena. Science only replies, like that scientist who last year replied when asked the question by a journalist, "Why was this year so rainy?" "Because it rained a lot!"

"Generally speaking, we do not consider the importance of science sufficiently and the importance of the reply to this question, "*Why*?"

The individual who calculates the speed, size, distance, weight, etc. of a star is easily thought to be unbalanced, and the learned misanthropist in the depths of a laboratory seeking the analysis or synthesis of a body in chemical reactions seems a madman.

"Let us not forget that the ink with which these lines are traced and the paper upon which these thoughts are printed are the result of science, and that the misanthropist who dimly sought reactions discovered the color that tints the feather of your hat, madame!

"Such are the results of the "infant science" of our time, and now you can see the practical results that the answer to this eternal "*Why*?" could give.

"Science is the soul of our existence, the generative impulse of humanity. If it were not for the slow progress of science, we would be but primitive beings clad in animal skins.

"This is one of the reasons why, in order to give humanity the means to facilitate its evolution, we must consecrate ourselves to the search for the ONE truth that is the answer to the question "Why?"

"Yet our science must not be exclusive.

"We must guide our researches with the means that the sages have given to us.

"We must not deny the intervention of beings superior to us mortals, since such intervention is proven.

"We must enter deeply into the truth, but not in order to use the results for our own personal well-being or our own interests or to satisfy our passions."

So wrote René Schwaller in an article that appeared in *Le Théosophe* on October 16, 1913. This article was the first of a series of seven that he wrote on the general theme of "The Value of Science." In his second article, "On the Atom," he stated:

"Physics tells us that electricity is an energy. But what is an energy? It is what results from work and everything that can produce work.

"To produce work requires the intermediary of a body. The result of a movement of this body will therefore be an energy and this energy that results from movement can produce work in its turn.

"Energy is hence something that results directly from an essential quality of bodies: inertia. According to this explanation, energy is something immanent in matter.

"But then *how can the electric atom be what produces matter when it is this matter that produces energy*?

"Of course scientists are well aware of this state of affairs, but despite everything they hang onto rationalism that led science on this path taking it away from truth. One would have to deny the work of a whole century if one were to confess openly what is being felt in the inner fortress of more than one learned consciousness. . . ."

In the third article, "On Light, Electricity, and Magnetism," Schwaller turned again to the crux of the problem, the question of the "Why?" or the cause:

"If we admit that the character of the cause becomes the character of the effects of this cause, we may conclude that different effects presenting analogies of character are children of a cause possessing this same special character. If matter is ponderable, all effects of this ponderability should lead us to take matter as cause. We may therefore state right away the following axiom:

"There is a constant analogy between the different effects of a single cause.

"As I said earlier, science travels from the effects to the cause: from material effects to their cause, which is matter. This is a deviated path—for should we not first know matter and all its characteristics before seeking to know the effects of matter? Our senses at present derive from matter, are the children of matter, and hence cannot conceive of anything but the effects of matter. If matter is an illusion, our senses exist only in relation to this illusion. The path from cause to effect is extremely simple and direct, and the results are extraordinarily simple, as I have myself been able to ascertain. There exist many phenomena of which science today does not even dream!

"If one puts the cart before the horse, one can certainly move forward, but if one puts the horse before the cart, one can surely move forward much more rapidly.

"Let us then work normally."

Every text must be read in context.

René Schwaller was born in 1887 in Alsace. His original tongue was as much German as French. All his life he wrestled with this inheritance, continuing, as it were, to "think in German" but write in French. This explains both the convoluted style of his writing and the philosophical bent of his mind: German being the philosophical language *par excellence*, providing the thinker with a flexible means of reaching the limits of rationality but no means of passing beyond them. All his life Schwaller suffered from the consequences of this linguistic destiny: Cartesian rationality and Hegelian depth. He applied himself relentlessly to the apparently impossible task of using reason to surmount reason. On his deathbed (as reported by his wife, Isha) he spoke of this struggle: "That one there gave me a terrible time!. . . Now I see his game: he was the enemy all my life. . . . I never thought him so fierce!. . . I knew that he was the obstacle, but I was unable to identify his tricks, the forms he could assume to deflect me from the path. . . It is him, you see, him, the mental being, who creates fear. . . and all our doubts, and all our dread. . ."

His father was a pharmacist (a chemist) and the young Schwaller passed his early years "dreaming in the forests, painting, and conducting chemical experiments in his father's laboratory." Thinking, art, and embodied activity or science thus divided his time.

At the age of 7 (1894), the boy received an illumination regarding the nature of the divine. We may therefore assume that he was no materialist; more than that, he had faculties, unusual cognitive powers. At 14 (1901), a similar illumination was given him regarding the nature of matter. Thus he found himself poised, as consciousness, the mediating term, between two absolutes, God and matter. These were his questions; he himself was the answer. Soon thereafter—perhaps two or three years, the exact date is unknown—he left home, a teenager, to go to Paris to continue his search. He studied painting under Matisse, himself then under the influence of the philosopher Bergson. Painting is important both because it is an art, a practical occupation, and because it has to do with, indeed depends on, perception, which is the essence of science also. What, after all, constitutes the proof of a scientific experiment? One *sees* what happens.

In Schwaller, then, we have a young man whose interests were equally religious, artistic, and scientific—three aspects of reality that were *one* for him, any theological or metaphysical insight requiring perceptual, i.e., phenomenal confirmation, the realization of which was an artistic question of the appropriate gesture.

More must be said. For the young man who arrived in Paris, these interests—metaphysics, art, science—were not the enthusiasms or pursuits of an ordinary student: they were the expression both of an innate knowledge, itself the fruit of previous lives, and of work and accomplishment in the present moment, i.e., an original, ongoing revelation. Innate knowledge prepared the field. Days and nights of prolonged study, chemical experiment, thought, and prayer—not to mention the practice of painting—opened the way to new insights. Alchemy was read and practiced; likewise the stonework of the ancient cathedrals was deciphered and interpreted. Above all, perhaps, H.P.B.—Madame Blavatsky—was taken as the avatar of the age.

From statements that Schwaller made later on, we may infer that the inherited path concerned the nature of matter, its existence in relation to consciousness and perception—this is the alchemical or Hermetic aspect—while the revelation, the present work, had to do with *measure*, which we may call the Pythagorean aspect. Hermetic and Pythagorean: two aspects of a single coin, like God and the matter that is the world.

Measure is the means by which matter becomes manifest as form. Such matter, generally speaking, is "potency" and exists as "substance" before all worlds and manifested states, awaiting only to be cognized to exist. Etymologically, as Guenon points out, matter is related both to *mater*, mother, and *metiri* (Skt. *matra*), to measure. It is what is measured, measure in this sense being a determination of potentiality. Measure is thus intimately related, on the one hand, to number (whose expression measure is) and, on the other hand, to order or harmony (Skt. *rta*, from which we derive our word ritual, and which is also related to the Greek *rheo*, to flow, as in the Heraclitean *panta rhei*, everything flows). Measure, then, is the order-producing gesture, which is a vision or illumination, as the *Fiat lux*, "Let there be Light," proclaims.

By the mere fact of being, therefore, any entity whatsoever manifests number, that is, the living relation of cause to effect.

In 1913–14, a member of the Theosophical Society, and by that evidence active in the occult circles of his time, Schwaller wrote a series of articles, quotions from which opened this introduction. His chief concern in these was to question the fundamental premise of science: its materialism. From another point of view, therefore, it was to bring science into relation with sacred science or esoteric teaching. This is theosophical agenda. Science is the soul of our existence, the generative impulse of the evolution of culture and consciousness, yet it does not understand what it is working with. The true nature of causality remains hidden from it. It knows only mechanical, horizontal cause and effect. But phenomena cannot be the explanation of phenomena and to simply gather and classify observations does not constitute knowledge. Science is but the embryo of science because it stubbornly sticks to the level of phenomena ordinarily perceptible and admits neither of a vertical axis nor of the existence of higher states of consciousness: higher cognitions, a higher phenomenalism. It lacks real explanatory power, because it will not rise above material phenomena: therefore it is helpless before the "Why?"

The *Study of Numbers*, published in 1917 and translated here, provides an alternative approach. This small treatise in the tradition of Theon of Smyrna, Nicomachus of Gerasa, and Iamblichus is the fruit of much thought and meditation and deserves serious consideration on the part of the reader. It cannot be read for information, but must be "thought along with."

The Ikhwan al-Safa write:

> "The philosophers have put the study of the science of numbers before the rest of the abstract sciences, because this science is strongly embedded in the essence of all of them. Indeed it is necessary for a man to reflect deeply and sufficiently on it, without taking an example from another science, but rather taking an example for other scientific subjects from it."

Nicomachus agrees:

> "Which of the four methods (Music, Arithmetic, Geometry, Astronomy) must we learn first? Evidently, the one that naturally exists before them all is superior and takes the place of origin and root and, as it were, of mother to the others. And this is Arithmetic, not solely because we said that it existed before all the others in the mind of the creating God, like some universal and exemplary plan, relied upon as a design and archetypal example by the creator of the universe to set his material creations in order and make them attain to their proper ends; but also because it is naturally prior in birth, inasmuch as it abolishes other sciences along with itself, but is not abolished together with them."

Clearly this is a youthful work. The more mature Schwaller de Lubicz would not have accepted everything the young René Schwaller wrote. On some points he changed his mind. But I think that generally speaking he would stand by most of what is written here. For students of his work it is of inestimable value to be able to catch a glimpse of this master mind in its formation.

CHRISTOPHER BAMFORD

Translator's Note*

René Schwaller in fact left emphatic instructions that this essay on the study of numbers, written early in his career, should not be republished without a note explaining that the passages concerning point, line, plane, and volume must be corrected in the light of his subsequent research into Egyptian mathematics.

Schwaller came to believe that the metaphor of creation as imaged by the point engendering a line, the line engendering a surface, and a surface engendering a volume was an arbitrary, misleading adaptation of Greek thought (Euclid and Aristotle). In contrast to this view, he presented his interpretation that the entirety of Egyptian philosophical mathematics was based on the notion that the primary expression of being is not the philosophical "point" but the three-dimensional volume. Hence he wrote in *Le Temple de l'Homme*, "Everything that exists is a volume. . . therefore a point is the apex of a volume, a line is the edge of a volume, and a surface is the face of a

*Based on conversations between Robert Lawlor and Lucie Lamy.

volume." These three components (point, line, and plane), when considered by themselves, are abstract concepts expressive only of mental ideation with no basis in the three-dimensional world of physical embodiment. By accepting the "original state" as volumetric, the physical world, which is also exclusively three-dimensional, then reflects the essential nature of its creative origins. Saint Bernard restated this Egyptian philosophical theme: "What is God? God is length, width, and depth."

Schwaller considered, furthermore, that Greek philosophy, in adopting the image of "Original Being" as point-like rather than volume-like, had helped to "deviate" the primary paradigms of Western thought toward the reductionist and mechanist assumptions that so plague modern man's relations to the natural world. All our thinking, designing, and engineering practices follow this model of using two-dimensional diagrams of points, lines, and planes, which are then superimposed on to the physical world of volume. But the reality of the natural world is exclusively three-dimensional, and, like a volume, inseparably binds opposed states: surface and interior; front and back; inside and outside; and top and bottom.

Much of Schwaller's subsequent work demonstrates the adherence of the ancient Egyptian mind to a volumetric model of reality, creating a "technique of thought" and a mathematics of an entirely different character and directive than our present ones derived from Euclidian and Aristotelian forms. This difference, Schwaller felt, has lead Western mathematics, science, and philosophy in the direction of predominantly abstract mental games, lacking the vitality and paradox necessary to create a harmonic dialogue between mind and nature. For these reasons, Schwaller warns against the "error of youth" contained in this otherwise insightful and informative exposition.

A STUDY OF NUMBERS

A Guide to the Constant Creation of the Universe

Foreword

Whenever our spirit or mind wishes to sift out from among the chaos of cosmic phenomena the truth, or at least the most "likely" reason for the being of things and their life, it needs a guide.

This need to simplify the world's appearance, to reduce it to a simple expression, may be the fact of our inability to extend our view beyond a certain limited horizon, an inability resulting from the imperfection of our sensory organization.

Whether this is so, or whether the world is really of so disorganized a complexity that it cannot be understood in its totality, amounts to the same thing. In any case, irrespective of whether or not the idea is accepted by people uninstructed in occult teachings, our sensory organization clearly seems to be imperfect. It is therefore capable of being perfected, though not by a "sensitive" completion of the senses, the *sensory memory*, or the natural mnemonic functions, but by a *perfecting of consciousness*. This latter requires the determination of "reactives" (senses), corresponding to the energetic activities and influences of the environment.

We lack direct consciousness of Space and Time. We can only know them indirectly by means of mass, force, and energy, and by the intermediary of phenomena which may be tested by one or another of our *five senses*. Human beings thus lack the two senses necessary for a knowledge of all causes. From this imperfection, of which we are always being made aware, the need is born to simplify. By this need everything is reduced to fundamental properties, without any attention ever being paid to the form of all the various effects of this universal organization. The result is that the science of numbers, the most wonderful guide to the continual creation of the universe, remains an enormous hypothesis. It will remain so as long as its use has not awakened in us the higher consciousness that usually escapes us, as long as we have not, by a deepened knowledge of things and their becoming, come to recognize numbers as truth, and as long, finally, as we have not experienced with our senses that the *living relation* of a cause to an effect is *truer and more real* than the effect could ever be.

Between a hypothesis and the truth exists a world; this world is the battlefield of reason and "emotion," which we define as the *pure sensibility* of the senses—an abstraction formed from actual sensation. In this world the "logical reverie" of the scientist and the ecstasy of the mystic meet; the first is analytic, the second synthetic, and both lead to the recognition of the science of numbers as the science of the basic laws of the universe, the science which fixes the proportions of the building, indicating the position of each stone and determining its moment of construction or destruction: *the Architect's plan*.

That there have been men who knew how to read this plan cannot—without doubting all history—be doubted. To cite but two examples, Plato testifies to the existence of a Pythagorean science, and Judaism attests to the truth of the Kabbalah.

To undertake the study of numbers in a fruitful manner, we believe it to be necessary to adopt the following general plan of study.

The five essential points, the basis of the study, will be observed in the following order:

1. Numbers, values and relations
2. The disengagement of numbers
3. The harmonic basis of numbers
4. The development of values
5. The establishment of harmony

1. Numbers are expressed by the figures 0, 1, 2, 3, and so on, up to 10. Here one will note immediately the double nature attached to numbers. There are, to begin with, numbers in themselves, forming a qualitative relation between each other. This is the relation between unity and multiplicity, with a fixed quantity of degrees and variation. Following it, there is the quantitative relation which results from counting things and defining a quantitative relation between them.

In this qualitative and quantitative function we may discern both the *nature of numbers*, their immanent, abstract life, and the *value of numbers*, their manifested, concrete life.

By the "abstract nature of numbers," I mean the *vital*

bond that exists between things. By the "concrete nature of numbers," I mean the *manifestation of life* under its many material, accidental aspects: weight, density, color, etc.

These two aspects of the nature of numbers have a common function: succession, by which the past, the present, and simultaneity, as well as the future, are defined.

Everything, in all things, may therefore be traced back to numbers, which are the last (or first) manifestation of matter, and the first cause of the creative *idea*. By this fact numbers are but the ideal and concrete relationship in the universe. Hence they constitute the principle of life, the vital impulse of the cosmos.

2. To understand true succession in creation, one must know how the first, or abstract, nature of numbers develops—how multiplicity disengages itself from Unity.

It is obvious that the first Unity, the cause without a cause, is indivisible. There are not yet any halves, thirds, etc. It is the first Unity. Hence it is purely qualitative, without quantity.

This first Unity is always, although under different expressions, the idea of the absolute, of the eternal, of the indefinite. This idea contains contraries (i.e., the same nature twice, but opposed in its tendencies), because the idea of an Absolute can only exist as the perfect stabilization of two essentially contrary natures. This stabilization cannot, however, exist, since manifestation immediately follows from it. In the last analysis, it is this idea that is generally meant by the term "cause without a cause."

This double nature of the first, abstract One is the

reason for the disengagement of multiplicity from Unity, as we observe it in nature, in each branch of a tree, or, indeed, in any natural phenomenon whatsoever.

Nature possesses, in itself, the tendency to make "the definite out of the indefinite."

The first One can therefore only create a multiplicity by qualitative addition, and never by multiplication, because multiplication is proper to procreation.

It is in this way, then, that 1 gives 1 and 1 or

1
1 1

and is by this fact, three. The indivisible One makes the first divisible number. This number, in its abstract nature, is 2 and becomes 1 as a concrete and *divisible* unity. Such is the *triple nature of the Creator* and this Creator is One, but a One manifested. According to the mystical explanation of numbers, the Creator God is referred to as this One, because in it one finds the father, the son, and the spirit: the creative *principle*, the created son, and the spirit which binds them together.

This is how numbers disengage from the abstract One.

3. The common function which determines past, present, and future decomposes into these three times beginning from the moment that the first One—the first, *indivisible*, purely qualitative, purely abstract cause—distinguishes itself into the nine other numbers that thereafter will constantly accompany it. This first cause

has potentially in it all future causes. Hence it presents another state, simultaneity, that comprises past, present, and future in a single Absolute. This is the fourth time.

By the coordination of these times, as well as by the diversity of the double nature of number, one obtains harmonies and dissonances, pure and mixed colors, and whole and fractioned weights.

This is the first reason for cosmic harmony.

4. This harmony is manifested in the complementary arrangement of harmonies and dissonances. This mutual "complementation" of two natures gives birth to new unities, which are then complex and whose base is abstract Unity. These new unities will be the origin of manifested numbers, and their quantitative nature. In this way, *values* develop.

5. Harmony can only reign in the world if multiplicity disengages itself from manifested, hence divisible, Unity.

This function is the same as that by which multiplicity disengages itself from the abstract One, but the act is now complicated by the fact of the preceding creation. What is creation in the abstract becomes the *first procreation* in the formative idea. This idea again procreates a second time; and then the concrete world is manifested, because only in it can what is procreated procreate in its turn.

Chapter One

The Irreducible One

The plan for the study of numbers that we have just described specifies what course should be followed in order to elucidate numbers from the harmonic point of view; namely, numbers are to be regarded as the philosophical basis of cosmic genesis.

This is obviously the deepest, but also, and perhaps for this very reason, the least "useful" form of this science.

Numbers possess a practical value apart from their high philosophical importance. Precisely this last characteristic gave birth to the mysteries whose veil, impenetrable to the profane, always enveloped this science—mysteries which can be known by anyone who takes the trouble to study, above all else, the metaphysical aspect of numbers.

The reason for the secrecy concerning the true nature of this science will then only seem "bizarre," incomprehensible, even incoherent, to those who like to learn because everything is explained to them and the teacher does the work that his students ought to do.

It has always been said that initiation occurs "by and in itself." One cannot *explain* the life of things; one can only merge oneself with it and thus *feel* it. In every epoch, therefore, the goal of all initiatory institutions was to give to whoever asked for it the *means of self-initiation*. Among those called in this way were sometimes found those who were chosen.

Numbers are the purest expression of the truth because they determine the exact relation between cause and effect. They allow one to know all the "hierarchical" functions which, from the cause, *give birth* to the effect.

Numbers, however, must not be thought of as a simple instrument of divination; this is only, one could say, a popular virtue given to this science by the ignorant.

Obviously, whoever knows this science perfectly will be able to foresee many phenomena, for he will understand the necessity and characteristic form of their evolution. In astrology, for instance, numbers possess a value unsuspected by the uninitiated. One must not, however, attribute to them virtues other than those which are their *raison d'être*: the proportions and hierarchical classification of the poles and relationships that bind the effect to the cause.

Doubtless the future is an inevitable consequence of the causes of both past and present, but one would still have to know all the causes in order to know the effects that will constitute the future. With numbers one can easily define the time, movement, and force which separate an immediate effect from its cause, but one must *know* (in the full sense of the word), not the physical, but the "occult" cause. This is humanly impossible, and only the "superhuman" being who has succeeded in merging with *space, the only quality proper to all things*, can know the occult cause of anything.

By numbers one can specify the dates (duration, in relation to unity: day, year, lunar month, etc.) of cosmic genesis, both macrocosmic and microcosmic. Thus the initiate will know all the essential conditions necessary for the development (birth, life, and "death") of all things, not only the stars, minerals, plants, animals, and man, but also their "hierarchy," that is to say, their evolutionary arrangement into races and subraces. In this does the marvelous power of numbers consist; this is its "utilitarian" goal.

The most difficult and most important study, however, is that of numbers from the metaphysical point of view.

We shall begin by discerning that every phenomenon comes to realization in three stages, as the archaic symbolism of the circle—containing point, diameter, and cross—teaches us. We shall call these three stages the cycles of *polarization*, *ideation*, and *formation* respectively.

The first stage, the cycle of *polarization*, is characterized by "generic selection," that is, the causal circle. A circuit of energy, for instance, tends toward a solution of continuity, a tendency that will find its satisfactory resolution in the most perfect opposition: its *complement*.

This function is of essential importance and must be well understood before continuing the study.

In order to completely understand this first polarization, we shall picture the initial circuit as a circuit of electrical energy on an uninterrupted conductor:

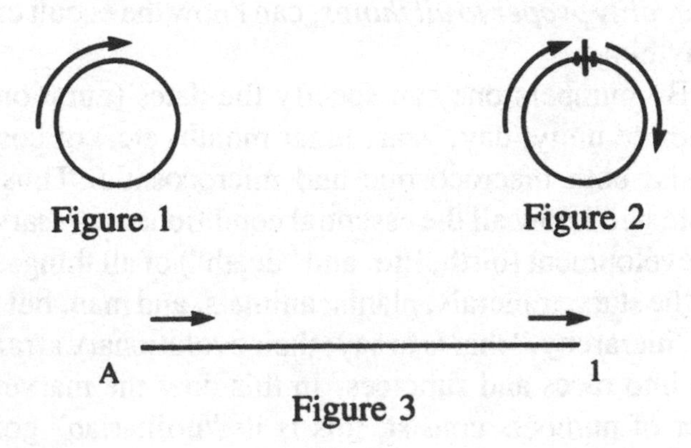

Figure 1 Figure 2

Figure 3

Figure 1 represents a closed circuit through which a current of electrical energy passes. The direction of this current is indicated by the arrow.

This circuit has no solution of continuity. It is therefore an "absolute," *nonmanifest* state. Only by virtue of our reason can we suppose the existence of such a state. This circuit is therefore ideal; our five senses cannot conceive it.

As soon as we become conscious of this ideal circuit, it becomes what Figure 2 represents, that is, a circuit with a solution of continuity created by the interruption at A.

Now this moment specifies for a determined *quantity* of energy *two qualities* of the circuit because the interruption of the circuit provokes both a moment of arrival and a moment of departure for the current.

Consequently two poles and a line of force result from it. The pole of arrival will be point 1 and the pole of departure will be point A (Figure 3). The first will be called the positive pole, the second the negative pole.

What provokes this interruption is necessarily a resistance. The reason for this resistance is the necessity of determining a quantity which will complete the unique and indefinite quality of the current.

The *absolute state* of anything is the abstraction of all quantity or divisibility from a thing, leaving only an indeterminate *quality*.

Since our consciousness is based solely upon the relationship of quantities to the Absolute, we can only conceive of this absolute state in relation to quantity.

By the opposition of *resistance* (quantity) to *activity* (pure quality) the phenomenon appears. The phenomenon is perfect when resistance and activity are in a *harmonic* state, i.e., at the moment when the resistance is equal to the activity. This moment cannot be indefinite, because the phenomenon, as the cause of its existence, possesses precisely *what is definite*: quantity. It will be limited by time and mass. The phenomenon is therefore again a cause working against a resistance which will be greater than the first, because, in involution, resistance or inertia increases in opposition to the activity which decreases. This resistance provokes a new phenomenon, and so on, until the establishment of a new harmony.

Eternal creation is thus the result of the opposition of a quantity to indefinite quality, and the size of this quantity is always momentarily equal to that of the quality which gives birth to it. By this fact the phenomenon is the result of the "complementation" of an active and indefinite absolute state with a passive (quantitative), definite state.

Each harmonic moment in the universe possesses by this fact its active cause and its passive cause, which is its complement. The "complementation" of these two poles must therefore have as its effect a new absolute, relative state, the new cause of the next effect.

Generic selection thus consists in the choice by the first cause (activity) of a corresponding resistance.

In the cycle of *ideation*, the first polarity—which has necessarily *become the new cause*—is distinguished by "*energetic creation*." This means that in order to constitute *the root of its future form*, the first pole establishes lines of force whose quantity and quality vary with its nature.

Here, for the third time, a *cause* (whose effect will be definitive in the cycle of *formation*) or *idea* (a polar complex having become a unity) appropriates other similar unities in order to fix the form in three directions: height, depth, and size. Hence the characteristic of this cycle is *formative growth*.

To better understand these three cycles, let us examine an ordinary example: crystallization.

In a saturated salt solution, the instant of polarization occurs when there is a perfect equilibrium of the conditions for a perfect solution (degrees of warmth or some other) and of the tendency to *energetic creation* in the salt. This tendency may also vary according to the complexity of the nature of the salt. "Generic" choice then takes place; that is to say: the most satisfactory, hence complementary, ambient condition (e.g., the walls of the vessel, a foreign body, a crystal of the same salt artificially introduced into the liquid) functions as an energetic neutralization and hence determines the *place* of the first crystal's formation.

Polarization is the scission or the cessation of the reason for the existence of a state.

We find in it the explanation of the phenomenon in which a saturated salt solution deposits its crystals more rapidly if a crystal of the same salt (or of a salt of the same kind) is introduced into the solution. In certain cases, this is even the condition *sine qua non* of crystallization.

When the first pole is established, it realizes its own equilibrium by putting itself in energetic communication (through the medium in which it is formed) with other points of "resistance." Then the marvelous work of organizing and neutralizing the different poles begins. Centers of attraction, formed by neutralization, become the centers of whirlpools which form spirals—regular or irregular as the case may be—and so on until the polarization corresponding to the nature of the salt is completed.

Between these different "complementary" poles, lines of force or "energetic rivers" are established which specify the axes of the crystal to come.

At this point the energetic creation of the crystal begins. Following this, the salt molecules have only to fix themselves according to the axes that have been traced, and crystallization enters into the phase of the third cycle, that of formative growth.

Thus sea salt, for example, grows on three regular axes—that is, it is limited by six passive poles, equal among themselves—around an active (neutral), central pole, giving the form of a cube.

This genesis is the same for every "species" of creation. It is the *comprehensible* aspect of all procreation that occurs by *generic selection*, *energetic creation* (fecundation), in order to form itself in the period of *formative growth*.

We have permitted ourselves this digression in order to give the reader a better grasp of the importance of the three cycles of genesis. Their importance will become quite obvious in the course of this little study. One will understand the cause of the variations in "tendencies" and the different vital manifestations that form the "complexity" of the nature of the species. This nature resides specifically in the *function of numbers*, at the time of the "becoming" of these species.

This "becoming" or manifestation of species results from the harmonic and disharmonic succession of the functions and poles of an initial, quantitatively indefinable state. The first activity is provoked by a disharmony, whose cause lies in the opposition of a *definite quantity* to an *indefinite quality*, as we said above.

Any idea of a mass or form pertaining to this first quality must necessarily be abstracted—which goes against reason. On the other hand, it is precisely our reason which forces us to hypothesize the existence of such a state having a single indivisible and indefinable quality which contains all forms or quantities. This state we named absolute.

The absolute state is the *first number*: the irreducible One.

All manifestation is therefore initially a result of addition, and then of the multiplication of this first Unity. Then the quantities so defined organized themselves in new unities, which are the *causal unities* of other combinations but are not irreducible unities, and therefore present the different qualities and quantities forming all the variants of the intelligible cosmos.

When a state of disequilibrium exists, its two contrary, nonequilibrated poles pass, during the "vacillation" of phenomenal realization, from perfect disequilibrium to perfect equilibrium, and through all the stages and nuances that the greater universe as a whole also passes through. Any perfect (that is to say, harmonic) unity, because its causes are balanced in their mutual action and reaction, is reducible as a unity. It nevertheless constitutes, insofar as it is a harmonic state of a preceding creation, the point of departure for a new phenomenon. Its successive poles are identical reproductions of the various fundamental phenomena of the cosmos, such as *succession (qualitative relation)* and *proportion (quantitative relation)*.

The simultaneous function of these phenomena constitutes the organized life of plants, animals, and human beings.

There are, therefore, two cosmoses in a single universe: the created cosmos of the irreducible One, which is the macrocosm, and the created cosmos of the reducible and harmonic One, which is the microcosm.

The network of all these multiple functions and the innumerable phenomena resulting from them, of which the universe is the simultaneous manifestation, may be reduced to a fundamental expression which is the establishment of the different phases by means of numbers, proportion, and hierarchy. It is impossible for us to *understand* the universe without tracing back in this way all things to these quantities of indeterminate quality called *numbers*. If one does not derive the cosmos from these proportions, the result seems to us like one of those futuristic paintings in which the most diverse impressions, realized in a single instant in the consciousness of an artist, are fixed on a single canvas. Doubtless it represents the truth of the moment, but the universe developed successively, that is, two agents presided over manifestation—time and movement. We should therefore take them into account unless we wish to lose ourselves in the illusion of phenomena.

Thus we come to understand that to know numbers is to know the universe, and that in order to study them we must (1) disengage numbers from the irreducible One until the first perfect, reducible *unity* is established; (2) follow these two causal unities through the creative cycles of *polarization*, *ideation*, and *formation*.

After this, we can replace *the number One* by some phenomenal cause (light, electricity), follow it through all the stages of "becoming," borrow or *find* a name corresponding to each variant, and in this way slowly move toward the knowledge of the marvelous edifice that is Hindu philosophy.

Is this to say that Hindu philosophy is unique? Not at all. The terms, the names designating the principles (*devas*) are *neither of our language nor of our habits, and our evolution consists precisely in adapting our Western language and habits to the designation and understanding of these principles*.

For this, Eastern philosophy is and ought to be a welcome instructor, but its role should be limited to that of example only.

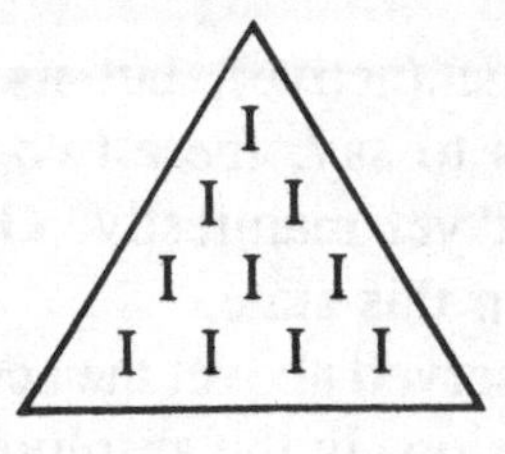

Chapter Two

The Cycle of Polarization[1]

Once the nature of the irreducible One is known, it is relatively easy to follow the disengagement of numbers according to the general scheme described above.

The irreducible One possesses a double nature, and this double nature, which is passive and active respectively, must manifest itself. Herein lies the mystery of numbers.

The cycle of polarization is characterized by generic selection. The irreducible One is equally feminine (pas-

[1] We have tried to make this part of the study of numbers, which is by its nature extremely abstruse, as clear as possible. In order to encourage personal research along these lines, it has entailed returning several times to the same phrases and giving points of reference in Kabbalistic studies. Given the small space at our disposal, personal research on the part of the reader will inevitably be necessary. We would have to quote entire chapters of Genesis, Plato, Theon of Smyrna, Saint Martin, etc., to be more explicit. We think, however, that it will be easy for the reader to orient himself with the help of intuition, for that sense is the most precious auxiliary in these advanced studies.

sive) and masculine (active), but its nature is not androgynous. That is to say, these two natures exist only potentially and not yet manifestly. Generic selection is hence impossible in this state.

But, as we observed above, the action of polarization is the fact of a scission. In the absolute state this scission has not yet occurred; hence the state is purely causal and has its being outside the cycles of manifestation.

In its absolute nature, the absolute state is not understandable. From the moment, however, that we observe its double nature we have already provoked the scission—for then One (the unique, irreducible cause) is recognized, by its double nature, to be Two.

Thus we may add a new Unity to the causal Unity, and this new Unity is Two.

I = Irreducible Unity. Absolute nature.
(II) = Reducible Unity. The manifested nature of the irreducible One.

Successively, then, we obtain Three distinct natures:

1. The irreducible One, active and passive in one. (Protoplasm)
2. Passive nature (feminine)
3. Active nature (masculine)

In actuality, the manifestation of the active nature necessarily precedes that of the passive nature, since the irreducible I, having to manifest itself, becomes II (the reducible I) and is above all 3 (III), and hence is the values

1 and 2, without which there would be no creation, as we realized in the previous chapter.

This fact is explained in the Bible in the words: God created man and made him a companion from one of his ribs so that he would not be alone.

Indeed, from the moment that there is activity, there is already opposition to it: *The infinite circle is broken; scission has provoked the first polarization.*

The first active nature which emerged from the irreducible One, or rather was created by It, potentially contains in itself the passive nature. The latter is part of the activity which, manifested either as active or passive nature, is from this moment on *androgynous*.

The first absolute number is father-mother. The second (polarized number) is androgynous. The third number is bisexual.

As it does among numbers, this function exists everywhere, creating the "races" or minerals, vegetables, animals, and man.

For the moment, however, the process of manifestation is far from being terminated. In the first polarized number we are still dealing with a created being and not yet with a procreated one. We observe two complementary natures, but they are still "interdependent upon each other." They must devolve to the point of being independent, that is, to the moment when the feminine and masculine natures become individualized and organized.

The created must have procreated and the procreated must likewise have procreated in its turn!

How are the two natures distinguished?

The active nature is what we called, in polarization,

the moment of the arrival of the current. In the circle it is *the part which possesses* and therefore the part which can give. It is quality, force, energy.

The passive nature is the moment of the departure of the current, the part which is left: *it is emptied* and hence can receive. It is opposition to activity, for it determines a quantity for a quality. It is the fecundated nature which creates by *the quantitative determination of vital activity.*

Quality is the number, and quantity is the value which measures and fixes the quality.

When the first three unities are defined and have formed the first two numbers, a new polarization occurs: a scission necessary for the manifestation of the two natures.

Considering the first triangle

I
I I

we find I, which is the father-mother, and II, which is androgynous. These two numbers form three unities.

Following in the sequence comes the number three or III.

The new triangle will therefore be:

I
I I
I I I

in which the first two numbers

I
I I

are creators. As such, they are a creative *unity* in manifestation and are different, therefore, from the absolute creative Unity.

The first triangle is composed of *two successive numbers* and gives us *three simultaneous unities*. Hence these three unities exist in the first triangle in a potential state and then become three unities *manifested by addition*.

From this moment on, the first triangle is a creative unity in relation to the third number. A man, similarly, becomes a father in relation to his child—a quality which cannot be taken away from him.

Creation is to be distinguished from procreation by the fact that it produces by an interior, hence purely qualitative disequilibrium a quantity which is the *formative receptacle* of the activity. Procreation, on the other hand, requires *an activity working on an independent resistance of the same evolution*.

If we consider the first triangle,

I
I I

we can see that it is composed of five natures: first, I, which is the two natures in One; second, II, which is twice I—such that in each I the two natures are potentially differentiated.

Hence this triangle has two passive or feminine natures, two active or masculine natures, plus a nature which cannot be defined other than by the very imperfect term "father-mother."

Thus, like the number three, *the number five is*

creator. But the number five still only constitutes the different nonmanifest natures of the three.

Creation will not be concluded until all these natures, existing in a potential state, are manifested one to one—that is, until they have taken on form, hence quantity.

In the beginning it is very difficult to disengage or isolate these different functions. As long as we are in a certain state, it is real for us, but, as soon as we pass to the state created by and hence following it, it is (apparently) no longer so. We descend with its creation, which is to say, our consciousness is modified. By this token the abstract is a real world so long as there is nothing concrete about it. As soon as there exists, however, a concrete world, our consciousness, appropriated to it, can no longer conceive the abstract by a concrete representation of or accommodation to it.

Thus one can see that II is the concrete form of I. These three I's, in their turn, produce a unity, which is *abstract* in relation to their own creation, which is III.

In other words, one cannot remove the idea of triplicity from that of unity, since it immediately gives us

I
I I

although in reality there are five natures in this triangle and 3 times 2, hence 6, in the number III.

Nevertheless in the number III each unity is a triplicity, giving us potentially *nine* "principles."

In the second triangle,

I
I I
I I I

we see the six natures of the III manifested to the *six unities*. With the five potential natures of the first triangle, these produce a complex of eleven new and creative "principal natures." Thus 1, 3, 5, and 11 are bound together, these being the creative numbers which, indicative of a creative and procreative disequilibrium, we find constantly in nature. These numbers fix the quality of organs which govern procreation with the aid of the two complementary natures of one or another of four realms.

The number III is a harmony in itself. It has three masculine natures and three feminine natures. It will only be able to *procreate* by the action of the first triangle working on it as fecundator.

The triangle

I
I I

is therefore One, undifferentiated; but there are three natures in this One—three creative possibilities. There, or in the irreducible One, we observe activity, passivity, and product in One: we distinguish three principles and we find in

I
I I

these three principles that have become *three natures*, giving a real unity, divisible by an even number (2) and an odd number (3), as well as by the absolute number, which is 1.

For this reason the number III is *Unity perfectly manifested*.

We had the creation of two by One; the first procreation of Three by Two; and now, in *Four*, we shall have *procreation by what has been procreated* by Three.

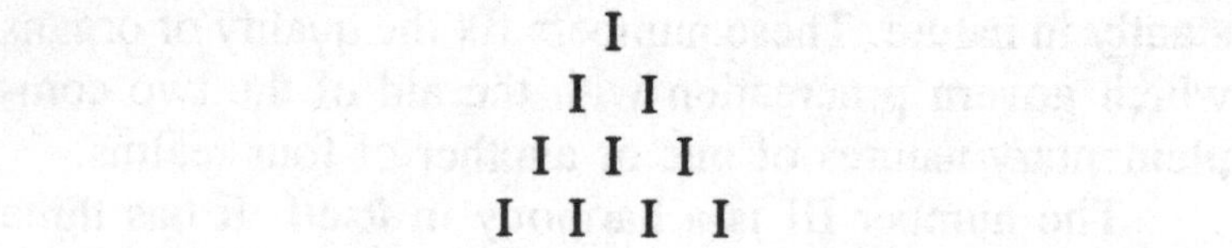

The IIII contains 4 times 2, hence eight natures, plus 1 times the triple nature of

I
I I

from which we derive the "potential" manifestation of the eleven natures of

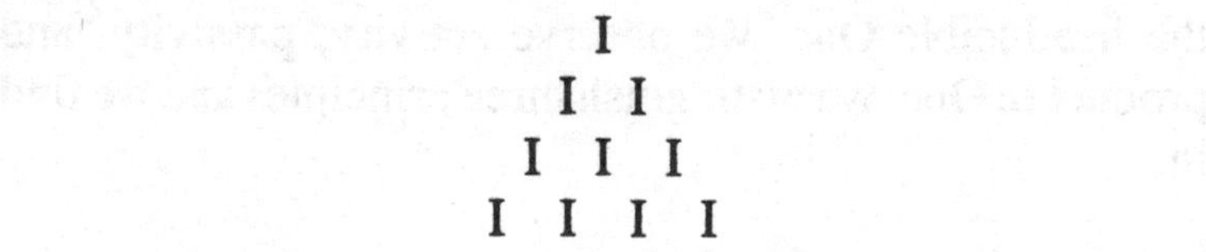

This quaternary triangle is the last. With it all natures are manifested. Since it has in itself the creative power

(11), it can procreate. As number it is 10, containing and surrounded by the nine principles: the irreducible One, the eternal fecundator.

This is the "Word" of Saint John, the "ineffable, incommunicable Word" of the Kabbalists.

It is the creative Spirit, the active principle that fecundates and maintains life.

It is the first power of the One, which means that the indefinitely great vitality of the irreducible One, once it is manifested as the first vital quality in 10, *becomes*—enters becoming. The first power of this 1 is 10 times 1.

The first "cause without a cause," I, produced by polarization the first point which, by addition, in its turn creates the line. Nevertheless, this point contains in itself the same faculties as any quantity of points. Hence if this point, an indefinitely small circle, can, in multiplicity, produce the phenomena of procreation, it can equally well do so by itself.

In the cycle of ideation we shall follow this phenomenon starting with 10, *which is the first point*!

We see that the triangle

I
I I
I I I
I I I I

includes 9 plus a unity. Now, 10 is always 9 plus 1. The number 10 exists only as a double number and, indeed, we cannot conceive it otherwise, since with 9 all the creative principles are manifested and 11 can only react upon a new, independent activity. Now (in relation to the Absolute), "procreation by what is procreated" equals a creation relative to phenomenal or concrete multiplicity. This triangle therefore contains in itself all that is necessary to multiply, and we rediscover in it the II of the triangle

I

I I

in which the II is the passive nature that immediately produces a new phenomenon by the activity of I. Now, 10 is 5 times 2, and we gave the quality of creating (by fecundating addition) to 2 and the quality of procreating for the first time to 5.

Here lies the basis of the occult calculations by which the duration and the relationships of the phenomena of nature may be specified.

By a similar function, Pythagoras established the well-known theorem according to which the sum of the squares of the two sides of a right triangle is equal to the square of the hypotenuse. These calculations were also used in the construction of the pyramids, which have a much greater importance than is usually given to them. In addition to serving as tombs for the pharaohs, they sheltered the mystery temples and were architectural books in which all the bases of occult calculations were given. In the Middle Ages, similarly, the cathedrals constituted, as it were, a golden book of secret teaching, wherein each stone is a page and each statue a profoundly symbolic initial, as each letter is the name of the Hebrew God.

Each unity is a triple *power* and a *double nature*,

active and passive. These two natures are linked to each other by mutual action and reaction, which explains the nature of "the Spirit between the Father and the Son"—the nature which manifests in the Two between the Three.

This double nature exists in every created thing, for there is nothing that is created if it is not by the definition of the two natures. And there is nothing that is not held in equilibrium.

Therefore the succession of number is, in order of appearance, 0, 1, 2, 3, 4, 5, 6, 7, 8, 9, 0: a series of nine numbers framed by two *zeros*, the first of which is a creative value, as the last is also, but the last is *the first power of the preceding one*.

The 0 that precedes the series of numbers from 1 to 9 represents the "absolute" 9, the first circle, "which limits the boundaries of the world"; it became 0 as the 10 came from the 1, but it belongs to the world of principles, to the causal world. Hence 0 is also a number, the most powerful of all, because it contains, as we may see by the evolution of 10, all knowledge and all powers.

In this concrete world we suffer under a constant illusion that links the end of the road that we see to the part that we have already traveled, and thereby we mix up past and future and make of it a present. The series 1 to 10 is therefore accurate only as "quality," because true succession makes us place one after the other: qualities, principles, functions, and quantities, from which follow quite another numerical order.

This is not the occasion to speak of it, but readers will probably notice how the numbers 31415 (which represent the famous π or the relation of the diameter of a circle to its circumference, the relation of 1 to 9) may be implied by the mutual, constant action and reaction of these two natures. In other words, nothing exists without the life of the Spirit.

Each thing in nature, reduced to its simplest physical form, becomes again divisible by 3 and by 2 and by 1.

From this we derive the alchemical theory of the *composition of matter*.

The way to compose this matter, to separate these two natures, therefore, becomes easy to know as soon as one knows the nature of its composition—as soon as one knows Numbers!

Up to now in our study we have learned to know numbers as principles and values.

From 0 we have come to 10, the first point, which is symbolized by the point in the circle.

As number (quality), 1 is abstract and becomes concrete only when it is 10, and 10 is 1 made concrete: *10 is the first power of 1*! The second power of 1 is 100.

Therefore, 10 (quality) is 1 (value) and the

I
I I

or abstract 2 contains its creative correspondence (in its passive nature) in 20, which is 2 plus 9 (for 0 always equals 9), and hence is 11.

The perfectly created number, the 3 of the causal or abstract cycle, is in manifestation 30 or 3 plus 9 equals 12!

That is why, in Kabbalah, the numerical correspondences of the Hebrew letters making up the sacred words are *whole values*, that is, powers of the creative qualities corresponding to the meaning accorded to these combinations of numbers and sounds (letters).

Continuing, we find for the procreative triangle (generative disequilibrium)

I
I I
I I I
I I I I

the value 40, which is 4 plus 9, or 13.

We find this celebrated number everywhere, in all religions and mythologies, and it always means a state of transformation. Superstition, which is moreover only superstition because it attaches a bad meaning to a truth, gives this number a capital and generally baneful importance. The truth is that this number manifests generative power. The effect will be good or bad, lucky or unlucky, following the *cause* and not a tradition.

We shall return later to this study of numbers beginning with 9 and shall then explain more clearly their becoming, their *raison d'être*, and their real value.

Now that we know the nature of the first abstract line (series of points), the first concrete point, that is, the series 1 to 9, we should learn to distinguish it.

In the first place, and as the always fundamental form, the two natures, masculine and feminine, active and passive, resulting from the relation of numbers to each other, separate out.

The numbers of a masculine nature are called odd,

those of a feminine nature, even. Their mutual action and reaction produce the whole cosmos, that is, all the ideas and all the forms.

The even numbers of our series are 1, 2, 4, 6, 8, then, in fact beyond the series, the number 10. They are even because they can be divided into two equal parts. They are therefore harmonic, which is to say, equilibrated, hence receptive, not creative by themselves.

The number 1 is also even, because it is, as we have seen, the father of numbers; having the value 10, it is even and odd at the same time. It is father-mother as value, just as the irreducible One is so as quality.

The odd numbers are 1, 3, 5, 7, 9. They are masculine because they are creative; they are active, disharmonic, hence disequilibrated. They cannot be divided into two equal parts and, except for the number 9 or 0, they can only be divided by themselves and by 1.

The number 9 is 0 manifested; it is its form, as we shall see later. The numbers 3, 5, and 7 are hence the creative principles of the world, the world having its existence by 9. Thus there is a triple world, a quintuple world, a septuple world, and these correspond to the cycles of polarization, ideation, and formation—for which lack of these numbers is respectively the foundation.

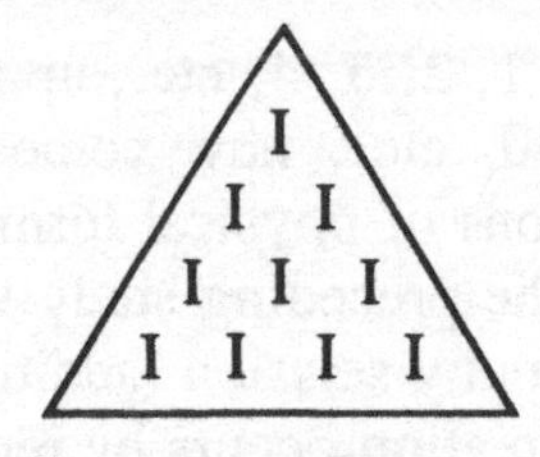

Chapter Three

The Cycle of Ideation

The idea is the energetic image of the form that it generates.

When the generators are selected in the phases of polarization, they arrange themselves in the phase of ideation so as to fix the energetic shape, the complex of lines of force determined by the generative poles, and to establish the "potential" skeleton of the physical form.[1]

In the preceding cycle we saw numbers develop from Unity and we found the numbers 1 and 3 to be the qualities attached to this cycle. The other numbers issue from it, but the numbers 1 and 3 alone are indissolubly linked to the cycle of polarization.

The phase of ideation in becoming (genesis) is essentially unstable and creative. It is the phase of the first procreation after the first procreation of absolute Unity.

[1] In this study the term "potential" ("power") is understood in the sense of "an accumulation of creative energy" and not in the mathematical sense of a function of exponents.

The numbers 1, 2, 3, 4, etc., up to 9, and then the powers 10, 20, 30, etc., now come to constitute the energetic foundations of physical forms.

Throughout the preceding study we have seen that first creation occurs by scission (addition), and we shall see now why procreation occurs by reproduction (multiplication) and how this happens.

The point, the first irreducible One, generated the lines by *addition*, for it cannot manifest itself by multiplication since 1 times 1 is always 1 and likewise 1 times a point always remains a point.

The point will not become a line except by being added to another point or, which amounts to the same thing, by separating one point into two.

Hence, in creation, addition and division have the same effect.

But addition presupposes the existence of an additionable quality, which can exist only outside the irreducible One.

Division, on the other hand, is the inverse function of multiplication. Hence the irreducible One must be assumed to be a multiplicity—and we now know how this original conflict is resolved: by the triple nature of the irreducible One, i.e., *the quality which allows one to conceive the procreating function, during which division is simultaneously addition—addition being the consequence of division, and division generating multiplication.*

I = the generative point
II = the point generated by division
III = the point generated by addition

The number III is the effect of I. The number II is generated by division, itself the inverse function of multiplication, *the cause not yet having an effect*.

The number III is the first complete solution of the creative functions, and, since it is engendered by the first multiplication of II, gives the first procreative state.

That is why the tetragrammaton, the tetraktys, and indeed all that archaic or magical symbolism expresses by the number 4, signifies both the last term of creation and the first term of the manifested existence of Being. Because it is the first stable figure, the square, 4 contains the secret of the first procreation.

This is the only phase of existence in which there is a first term which is created and a last term which is the first procreated one, characteristic of the phase of ideation. The point can only create by addition (division), and the resulting line can only procreate—and never create. A line added to itself is, and always remains, a line, whereas as soon as it is multiplied by itself, a line becomes a face.

The numbers 1, 3, 5, 7 are the creative numbers since they do not contain any procreated form. They are the prime or uncomposed numbers. In other words, they generate all figures and forms, and play, with regard to concrete manifestation, the same role that the irreducible One plays with regard to the abstract world.

By addition these creative numbers can produce other prime or uncomposed numbers (in the sense of procreation), such as 11, which is the first representation in the concrete world of the function of II in the abstract world. This is because 11 is intermediary, just as II is, and results from an *addition* of 1, 3, 5, and 7, and does not include multiplication.

The numbers 1, 3, 5, 7 are the generators in the phase of ideation. By multiplication with themselves they produce faces, are only engendered by addition, and provide the perfect transition from the abstract to the concrete through the phases of ideation.

We have said that the square is the first stable face. By stable we mean what is the consequence of an always equal function and produces no complex effect that diverges from the natural consequence of the first function.

The face which has four sides is not the first face, for three vertices are sufficient to establish a face, so that the first face is therefore the triangle. The triangle, however, as soon as it is multiplied by itself, immediately produces a square. It differs in this from the first stable face of the square, as we shall see in the next cycle.

The function presiding over the cycle of ideation is the multiplication of a simple or complex unity not itself engendered by multiplication. *The natural consequence* of this function is the creation of the face.

The triangle does not conform to this law. It is a face not engendered by multiplication but resulting from the first scission (addition).

On the other hand, it is not really a definitively constituted or stable face, for it *engenders* a stable face as a result of multiplying itself.

The first perfect triangle would therefore be one that would fulfill the following conditions: first, to be the first face; second, to be constituted by three vertices; third, to engender a stable face by multiplying itself; and fourth, to be constituted by a stable face or, at least, to contain such potentiality.

The apparently insurmountable difficulty that human reason encounters here is easily resolved by the knowledge that this short study of numbers has given us.

The first triangle

I
I I

fulfills all conditions except the first and the fourth. Indeed, there cannot be a face which only includes one unit of height to two units of length, for this constitutes a line and not a face. Neither does it contain a stable face as the fourth point requires.

The second triangle

I
I I
I I I

fulfills the first, second, and third conditions.

The fourth condition will therefore be met, beginning with the third triangle:

I
I I
I I I
I I I I

This constitutes the tetraktys or quaternary, which, first, is the first possible face; second, is triangular and, by this fact, made up of three vertices; third, engenders by multiplication with itself a stable face, which is *9*, as was shown above; and fourth, potentially contains the first square, which is *4*.

Since the first triangle

I

I I

is, as we know, considered to be the first reducible unity, we must regard the quaternary triangle as being potentially the first triangle constituted by three equal angles, which we can show thus:

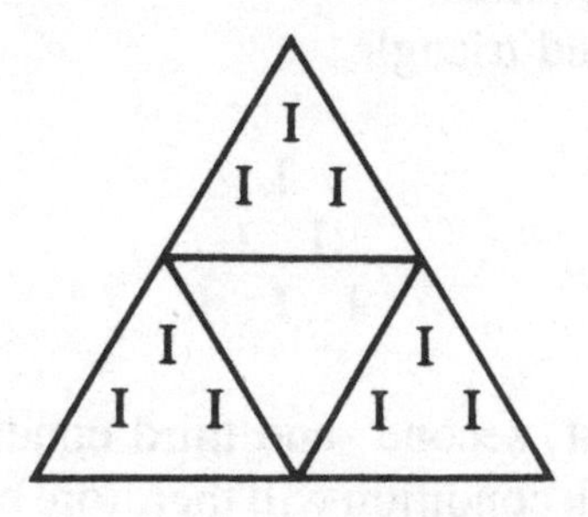

This determines a triangle whose height is half as great as its base. This triangle clearly frames absolute Unity, which is the vital breath in the things of this world. There are therefore in the quaternary triangle actually two distinct triangles. These are the triple creative nature in the One and this triple nature as manifested in the thrice three unities of the first triangle.

This figure, which is the first manifest triangle, fulfilling all the conditions laid down above, engenders the perfect square, which is 9, for it contains all the numbers.

In this way all squares become divisible into two triangles, since 4 is 1 and 3; 9 is 3 and 6; etc. The squares, beginning with 10, become compounds.

The uneven (positive, masculine) numbers form the squares of whose two triangles one is negative and the other positive (even and uneven). The even numbers, meanwhile, give the squares whose triangles are both negative or both positive.

We cannot enter here into any explanation of the reason for this fact. It will be sufficient to indicate to the student that the even numbers are all compounds: 2 equals 2 times 1, 4 equals 2 times 2, 6 equals 2 times 3, 8 equals 2 times 4, etc., and this function determines the intermediate or variant forms of the square, such as the figures or the parallelograms with two unequal sides.

We have said that the essentially creative numbers are 1, 3, 5, 7 and that they establish the following principles which we find everywhere at the base of creation of the two natures.

The multiplication of any quantity by itself (procreation) results from the addition of a positive, masculine, odd nature to a negative, feminine, even nature.

The other figures—pentagons, hexagons, polygons—are the effects of an addition of different triangular and square figures, just as the numbers above 10 result from the fact of the addition to this 10 of the numbers 1 to 9.

The square is the stable base of forms.[2]

[2] The student may consult Theon of Smyrna for the "technique" of numbers. To develop the latter here would take us too far afield, given the small compass of our study.

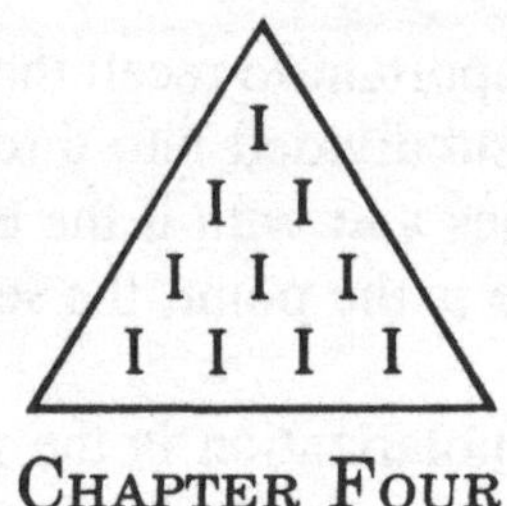

CHAPTER FOUR

The Cycle of Formation

Addition of the point gave the line. Multiplication of the line created the face. And by multiplication, similarly, the face will give the form. This constitutes the cycle of formation.

The square is the perfect face. The perfect form, then, will be that which results from the multiplication of the square: the cube.

As in the preceding cycles, the face, in order to become the cube, will pass through all the phases of polarization and ideation. The square again becomes the point—the point of departure—for the form. The first pole for the creation of the cube, or of the solid generally, is the face. The cycle of the polarization of the solid therefore begins with the face, which is one-dimensional because it possesses a spatial tendency in one direction only, then proceeds to the second dimension in order to become the third dimension (three-dimensional). This last phase is *the cycle of formation in the cycle of formation.*

It is always important to recall that each of the three cycles is likewise subdivided into three cycles, and that only the cause varies and with it the effect as well.

The first cause is the point, the second the line, and the third the face.

The cycle of polarization is the phase of addition, whereby the nondimensional becomes the first dimension: the line.

By multiplication the cycle of ideation gives the second dimension, the face, and thus the face gives the solid by multiplication of a first multiple.

At the base of all creation is movement.

Addition is the first, or linear, movement.

The first multiplication is the second, or planar, movement, and multiplication of the latter gives the third, or solid, movement. In a solid, movement decomposes into these three *principial movements* if the body moves individually and not in relation to another body—in other words, in rotary movement about itself. Thus its first or linear movement constitutes the axis, and its second or planar movement the equator. Its third or solid movement results from the mutual action or reaction of the first two, of which one diminishes as the other increases, producing centrifugal force.

The symbol of the cross in the circle represents this constant game of the first two movements, which are even more strongly marked in the symbol of the swastika. As soon as there is a face, there is also a reason for the immediate formation of a form. There is a reason, similarly, for the existence of rotary movement as soon as there is a creature, or an Effect, from the line to the form, because at its base is always the existence of a formative axis or a first line of force.

Procreation is, to begin with, the function of a linear movement, and thereafter of two circular movements.

We have said that the first face is the triangle, whereas the perfect face is the square.

The first perfect triangle is

I
I I
I I I

and the first perfect square is given by the quaternary:

I
I I
I I I
I I I I

The fundamental or first perfect form will therefore be that which is made up of these two triangles, that is, of 16 units. Indeed, the first square, or 4, gives the first perfect form or cube, which is 16.

The triangle alone cannot give a form by the simple procreating function (i.e., multiplication), because a triangle times a triangle gives a face, which is 9.

The first form therefore comes into existence by, first, the addition of two triangles, which gives the square; and second, by the multiplication of this square.

In principle, the first triangle is I, the second being

I
I I

These two *added together* give IIII, or 4, and this multiplied gives 16, or the first cube.

But the cube is not the first *form*.

As the triangle is the first form, having an effect only by addition, and not at all by multiplication—as we saw above in the cycle of ideation—so the form has a first existence which cannot enjoy the procreating qualities of the perfect form. That is, there is a creative or causal form whose ideal form is the cube. This first form necessarily has a triangular base and is engendered by the addition of triangles.

This form is called the tetrahedron.

Four times the triangle gives the tetrahedron, but from the face of the establishment of the three sides of this form, the fourth (the base) exists. In *principle*, therefore, it is made up of 12 units; in *fact*, however, it is made up of 9. Nine is the cause of the final cube: it is the cause, but it is so only manifestly by 12. However, from the moment that the base is constituted by the fourth triangle, hence when it has established the twelfth point, it creates the reason for a form, whose multiplication by 12 gives the cube. It is made up of 24 times the tetrahedron whose base is formed by the triangle defined by the diagonal of the face of the cube and whose apex is the central point of the latter.

All forms may be derived by addition from the cube and the tetrahedron.

Beyond the third dimension, another dimension cannot exist, and what is usually called the fourth dimension is actually the first *nondimensional* state, *quality*, causal space, or, in absolute terms, the Nothing or *irreducible One*.

This is all that one can say in so brief a study of Numbers, or first proportions: the divine measure of the maker of the universe, intelligent as the creator, unintelligible to creatures, and the eternal cause of intelligence.

In summary, we may now apply to each of the three cycles the following qualities and functions delineated on the chart reproduced on the following page.

In general terms, and relative to the irreducible One (the eternal Creator), all creation is passive, and this first activity alone produces, through all forms of its manifestation, the Effects which are the cosmos.

In particular, the different effects of this constant creation are more or less passive in relation to each other, from which follows the relative activity of matter, and masculine activity as opposed to its feminine nature. Seven terms (factors) constitute the cosmos in appearance; 9 terms (functions) constitute it in fact. In the ideal succession indicated above, the term III exists as much as this term in I, etc., once it is Effect, it becomes cause: the Son partakes of the Father as much as he does of men, as the Son of God himself said.

We said that movement is the basis of all, because only procreation is the creative form of which we have sensible knowledge. Where for us mortal men, there seems to be creation, there is already procreation. Absolute Creation is not of our sensible world: only *reason* shows us its existence and, beyond reason, only mystic clairvoyance can *prove* this truth.

Procreative or circular movement is a function of

Cycle of Polarization	Fecundation of selected sympathetic natures: Addition. Creation.	The irreducible One becomes II. Creation of the five principial natures. First state of creative disequilibrium.
Cycle of Ideation	Energetic formation. Absorption of the seed by the fecundated nature and the establishment of lines of force. Resolution of the two natures into *one* new androgyne containing the two natures at the same time.	I I I becomes reducible I, the first quantity, the fundamental energetic form. The One fecundates the II and produces III, which is the reducible I, a nature equally active and passive.
Cycle of Formation	Formative growth. Internal scission of the two natures so that one predominates. Growth of the new seed in quantity, and *in form*, to maturity.	III becomes IIII. Function of multiplication, movement. II, which is 2 times active and 2 times passive, multiplies and becomes 4, the seed-face of the next form.

The succession of numbers, as proportion and measure of the cosmos, is the following:

I = activity, II = inert reaction, III = Effect.

Cycle of polarization. Creation.

III = I (reducible) new active cause, IIII = second inert reaction, IIIII = Effect.

Cycle of ideation. First procreation.
Multiplication in quantity.

IIIII = I active and I passive, hence I androgyne, a new active cause, IIIIII = third passive reaction, IIIIIII = Effect.

Cycle of formation. Second procreation.
Formative seed.

By scission this last cycle produces the two distinct natures which will procreate in what follows.

Time and Space. Time and Space, irreducible magnitudes, are the essential factors of which modern men of knowledge never take account, but which nevertheless are the most important, as simple observation of nature makes quite clear.

Everything in nature is therefore based upon and limited to the fundamental proportions that we have just sketched out in this study. If we are often to rise directly to the immediate cause of a phenomenon, we can, by observation and by analogy *based on these first proportions*, arrive at the inward nature of the phenomenon, its immediate cause, and at its own effects.

I think that we cannot do better than to end this modest little work by citing facts and phenomena that science has uncovered and that corroborate the Science of Numbers.

The proportion of the planets' distance from the sun is the ratio of 7 for Mercury, 14 for Venus, 21 for the Earth, 28 for Mars, 35 for Jupiter, and 42 for Saturn. This is a septenary cycle (with the sun in the middle) and constitutes a new unity for the most distant planets, whose proportional distances among themselves are again, 7, 14, 21, etc. In time, however, the Unity is no longer 7 but 14, and is disproportionate with the previous cycle. In occult science the existence of 12 planets is taught. Astronomy does not yet know all of these, which is why we do not note them here.

A revolution of the moon about the earth is fixed at 28 days, that is, at 4 phases of 7 days.

We now know that for us the fourth term is the creative, that is, the visible and sensible term. *One term* is given manifestly by a septenary system, that is, a succession of seven factors resulting from any activity on its corresponding resistance, and nine "occult" principles or functions.

It will not surprise the student, therefore, should he recall Darwin's observations which state that mammalian gestation is a function of a determined period of lunar phases. Thus pigeon eggs hatch in 2 lunar phases, chicken eggs in 3 phases, duck eggs in 4 phases, goose eggs in 5 phases, ostrich eggs in 7 phases, and human ova in 40 lunar phases of 10 lunar months or 9 solar months. In the light of the study set forth above and these facts, words become completely unnecessary. One may simply note that the human ovum is the most perfect of all organized realms actually living.

We limit this truth to what actually is, for it is certain that life transported to a sphere farther from the sun, that of Mars for example, would have to endure the consequences of the new proportion, that is, 5 times 7 instead of 4 times 7, which is ours. For us, everything is limited to the proportion of 4 times 7 or 4/7 of the total activity.

As we only perceive what is outside us and above us, just as the eye can only see what affects it and cannot see itself, it follows quite naturally that we are in the fifth Kingdom, since we can perfectly perceive all that is of the fourth. We possess five senses, of which one, touch, is common to all; it is the central sun, and the other senses (planets) are but modifications of this sense according to the mode of activity surrounding them.

These observations and others still more irrefutable are the reason why occult science says that we are living in the fifth subrace of the fifth great race of the world, and that a sixth and seventh sense, both already existing germinally, are in the process of developing. They will become perfect when the sixth and seventh subraces have arisen.

For these reasons also the quintuple world is as much of a truth as the septuple world—if one realizes the relativity of each.

One could multiply such examples establishing the perfect truth in this study by analyzing any phenomenon. We shall limit ourselves, however, to citing a physiological observation concerning cellular segmentation, the ovum, that is, concerning the function of numbers in the first form of organized life:

> In the beginning an organism is composed of *a single cell*, the ovum, of which we have already spoken, and whose *segmentation, as the model of the generation*, the proliferation of cells in general, we have already described. . . . As soon as the ovum is divided . . . [at first into two, then] into four segments, these segments already delimit by their separation a space between themselves, the so-called cavity of segmentation. In the ratio that the segmentation continues . . . [by 7 and 14] this cavity grows larger and larger, until finally the segmented egg forms a *hollow* sphere, whose wall is made up of a simple layer of cells comparable to an *epithelium*.[1]

[1] M. Duval, *Physiologie*, p. 14.

BOOKS OF RELATED INTEREST

The Temple in Man
Sacred Architecture and the Perfect Man
by R. A. Schwaller de Lubicz
Illustrated by Lucie Lamy

Sacred Science
The King of Pharaonic Theocracy
by R. A. Schwaller de Lubicz

Symbol and the Symbolic
Ancient Egypt, Science, and the Evolution of Consciousness
by R. A. Schwaller de Lubicz

The Temple of Man
by R. A. Schwaller de Lubicz

The Temples of Karnak
by R. A. Schwaller de Lubicz
Photographs by Georges and Valentine de Miré

How the World Is Made
The Story of Creation according to Sacred Geometry
by John Michell
With Allan Brown

The Golden Number
Pythagorean Rites and Rhythms
in the Development of Western Civilization
by Matila C. Ghyka
Introduction by Paul Valéry

The Dimensions of Paradise
Sacred Geometry, Ancient Science,
and the Heavenly Order on Earth
by John Michell

Inner Traditions • Bear & Company
P.O. Box 388
Rochester, VT 05767
1-800-246-8648
www.InnerTraditions.com

Or contact your local bookseller